Concre

Step by Step Guide to Making Your Own Diy Countertop

By
Bradley Specter

ISBN-13: 978-1541115040
ISBN-10: 154111504X

Table of contents

Initial Planning — 2
    Thickness of the slab — 3
    The colour — 4
    The dimensions of your concrete countertop — 5
    What type of sides do you prefer? — 5
    Do you prefer a mold or a cast-in-place? — 7

Materials you will need — 9
    Material for the frame — 9
    Something to mix concrete — 10

Making the mold — 11
    Setting up the frame — 11
    Attention to special cut-outs — 12
    Use silicone to make the edges smooth — 13
    Purpose of using olive oil — 14

Reinforcing the concrete — 15
    Why should you reinforce? — 15

Coloring the concrete — 19
    Using integral pigments — 19
    Using dyes — 21
    Using acid stains — 22
    Using decorative aggregates — 24

Mixing the concrete — 27
    Concrete mixes — 27

Pouring the concrete — 31
    Where to pour the concrete? — 31
    Pouring concrete on a mold — 31

Pouring concrete directly on top	31
Vibrating the mold to remove air	32
Screed the top	34
How to let the concrete slab dry slowly	35
Finishing the countertop	**37**
Ways to check if it's dry	37
Turning the countertop over	38
Protecting the countertop	**39**
Wet polishing	39
Penetrating sealer and concrete sealer	40
How many coats?	42
Is cutting food on countertops toxic?	43
Final words	**44**
Conclusion	44
Thank you	45

According to most people especially in the United States, owning a kitchen with smart countertops designed with concrete is a high priority. Since the kitchen is considered as the focal point where all culinary creativity takes place, a well-designed countertop will be key to filling out the equation.

In the past, people relied heavily on laminates to design kitchen countertops. Although these materials were exceptionally stylish and up to the task, they lacked some key features which forced most homeowners to research on suitable alternatives.

Due to this reasons, the concrete countertops were unveiled. Today, over 80% of United States homeowners rely on concrete countertops as a way of expressing their style, individuality and design preferences.

I believe most of you have been asking why all these hassles of installing concrete countertops while there are so many decent options available such as marble, glass, ceramics and stone tiles. To answer this question, let me say that there are so many benefits offered by concrete countertops that are not offered by other materials. Some of them include;

- **Durability**—unlike other stylish kitchen countertop materials, concrete countertops are strong, scratch proof and hard to chip off. This material also requires less maintenance making it perfect for do it yourself persons who don't enjoy frequent replacements time after time.

- **Heat resistance**—another benefit of concrete countertops is that they're heat resistant. This benefit has been praised by most kitchen enthusiasts because they don't have to worry placing hot pots, cooktops or pans on top of the countertop (keeping in mind your used water sealer).

- **Style**—the fact that concrete countertops can be designed with various colors makes them the best choice for stylish people because you only need to select a color that complements your kitchen décor.

Initial Planning

Now that I've given you a solid introduction to concrete countertops, let's now move to our first stage of creating the countertop which is the initial planning stage.

Making your own DIY concrete countertop is actually one of the most challenging but rewarding tasks in the DIY arena. To ensure that the entire project goes as planned, you'll need to set a solid foundation right from the beginning.

For beginners who've never done this before, the first step is to organize all the materials required for the entire project as well as take precise measurements of your kitchen space where the concrete countertop will finally rest. Perform a thorough research on the quality/quantity of concrete you'll need, the materials you'll need and the size/quality of the melamine board you'll require.

Thickness of the slab

Concrete countertops come in a variety of different thicknesses depending on your preference. However, according to most contractors, the standard thickness of most concrete countertops is either 1.5" or 2"(3.8 or 5cm).

Such a thickness is regarded as a standard concrete countertop thickness and is commonly used and due to some benefits. Firstly, thickness directly affects the weight of the countertop. With a thick countertop let's say 6"(15.25cm), it becomes very difficult to install (*due to its massive weight*) resulting to damages if not handled properly.

Secondly, the average thickness of 1.5" to 2" gives your kitchen's concrete countertop a streamlined appearance making it stylish.

Note that, the thickness of a concrete countertop is directly proportional to the weight. The thicker the countertop, the heavier it becomes. Due to this reason, it's good to understand that a 1½" (3.8cm) thick countertop will have an average weight of 18.75 pounds/sq foot (91.5kg/sq meter) which is the recommended weight for an average concrete countertop.

The thickness of a countertop

The colour

Concrete countertops are available in a variety of different colors depending with your taste and preference. Among the common types of are;

- Gray
- White
- Brown
- Custom Orange
- Custom Blue
- Custom Purple

Gray is the most preferred countertop color by most due to its availability and ability to keep the kitchen neat and natural. Other common colors which you can choose from include white and brown. While white is preferred due to its ability to keep the countertop clean, brown is also a common choice due to its natural appeal and ability to resist stains.

Finally, custom colors are also moderately popular especially to restaurants. The purpose of considering custom colors is to ensure that your concrete countertop complements with your restaurant's or home's kitchen décor.

The dimensions of your concrete countertop

After you've figured out about the colors, you now need to measure the dimensions of your concrete countertop. To do this, take a tape measure and carefully record the measurements of your existing kitchen countertop and the base cabinets. When measuring the cabinets, remember to add an extra ¾" (2cm) for the overhang (*this is important*).

After you've done that, the next step is to determine the length and the height of your countertop. According to statistics made by professionals, the average measurement of a normal kitchen countertop is usually 2' wide by 6' (0.6 by 1.8 meter) long. When it comes to the height, most kitchen countertops measure 36" to 38" (91 to 96cm).

However, if the countertop is intended for a bar or to function as a serving counter, then the height varies from 42" to 46" (106 to 116) tall (*these are measurements from the ground to the top of your concrete countertop*).

What type of sides do you prefer?

Adding decorations on the sides of concrete countertops is a brilliant way of making your kitchen or bathroom countertop stylish and elegant.

When it comes to applying the details, you need to figure out the type of sides you really want. There are different types of molds/forms a concrete countertop can be designed into ranging from rope edges, vines, stones and many more. To mold the sides, you can either use liquid rubber or liquid plastic (*mostly used by professionals*) or you can simply go for the cheaper melamine and wood.

However, let me say that, liquid rubber and liquid plastic may be expensive methods but they're relatively fast and precise when it comes to producing high quality well detailed edges/sides. On the other hand, melamine and wood is a cheap and affordable method but it takes some time to set and screw the frame to the desired shape. The melamine method is preferred by most people.

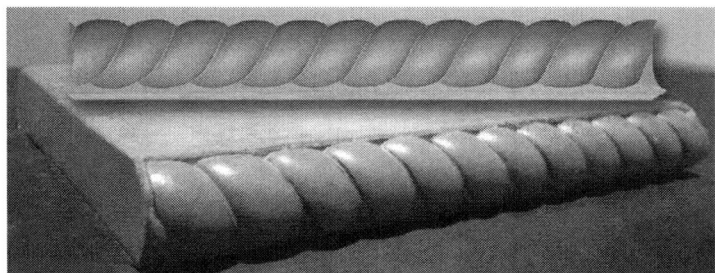

Rope edges (source: www.stonecretesystems.com)

Do you prefer a mold or a cast-in-place?

There are three major ways of creating a concrete countertop on your kitchen. Depending with your preference, you may decide to completely replace your kitchen countertop with a newly molded concrete slab, prepare a cast-in-place or you may decide to paint the top of your countertop with concrete. This is a decision you have to make before proceeding with the rest of the work because once the job is done, it becomes very difficult (*mostly impossible*) to reverse the decision.

- **Creating a mold**
  This method involves creating a completely new concrete mold which after curing will have to be carried and placed on top of the kitchen cabinet to form a concrete countertop.

- **Cast-in-place**
  Also referred to as pour in place, this method involves building forms on top of kitchen or counter cabinets to support the concrete that will be placed on top. The process of preparing concrete in this method is very similar to precast method. Cast-in-place method is mostly suited to large scale projects where precast would be almost impossible (*it would be too costly, time consuming and very risky to transport large concrete slabs*).

Cast in place countertop (image from www.concretecountertopsolutions.com)

- **Painting the countertop with concrete**
  The third method of preparing concrete countertops is by painting the top of your kitchen counter with wet concrete. This method is mostly suitable to homeowners who don't wish to remove the top layer of their kitchen countertops. In this book we will focus on the "creating a mold method".

Materials you will need

In this chapter, we will use example dimensions. You can change these to match your own needs. Keep in mind the subjects we talked about in the previous chapter.

Material for the frame

When making your own concrete countertop, always make sure that you assemble all the necessary materials and tools that you'll require when mixing, setting and molding your concrete. Although there are many materials one can rely on, I will highlight the best materials which are cheap and easily obtainable.

First, you'll need a melamine coated particle board measuring about 4' by 8' (1.2 by 2.4 cm) with a thickness of 1½" to 2" (3.8 to 2cm).

Secondly, you'll need additional melamine boards (or scraps from the first board) which will be screwed to support the sides of the mold. Remember that the side boards are responsible for determining the thickness of your countertop (because we will remove all the excess concrete that's above this height).

Something to mix concrete

After you've set the mold, the next step is to mix the concrete. There are various ways you can rely on when mixing the concrete such as using your hands, using a bucket or hiring a concrete mixer.

Using your hands
When using your hands, remember to wear waterproof gloves to avoid the cement from damaging your skin (*cement reacts with water and can damage your skin or eyes*). Place the concrete mixture in a wheelbarrow, add water then use your hands to mix it thoroughly and evenly. Keep mixing the concrete until all the dry pockets are gone.

Using a mixing machine
If you're using a mixer, the first thing is to ensure that all the materials needed for the operation are ready. Start the machine then begin by pouring sand, rocks and pigments. Add water, all admixtures, maybe fibers and pozzolans before adding cement gradually. Mix the ingredients thoroughly until you get the desired quality to suit your concrete countertop. When mixing the concrete, only add a little amount of water to make it strong and unable to crack during caring.

Using a bucket
One of the most common methods of mixing concrete when preparing a DIY concrete countertop is using a bucket and a manual/electric mixer.

Making the mold

Setting up the frame

Now that we have the materials ready, let's get started with setting up the frame. The first step when setting up your frame is taking the correct measurements of your kitchen countertop.

Repeat the measuring process at least twice or thrice to ensure that you have precise readings. Set your board firmly on a pair of sawhorses then Measure and mark the exact dimensions on your melamine board.

Using a circular saw, a jigsaw, a table saw or a miter saw (*they are all perfect for cutting melamine boards*), cut firmly to get your desired shape. When you're done, cut the sides of the mold according to the measurements you've taken.

If you want your countertop to overhang (which you should), you can add an extra ¾" (2cm) on each side of your board. Keep in mind your walls, kitchen cabinets or fridge because you don't need an overhang there!

Attention to special cut-outs

In such a scenario, simply measure the dimensions of your sink/cooktop then mark them on your melamine board. Measure very precise the distance from every side of the countertop to your sink of cooktop. Then use some scrap melamine board to make a square or rectangle (according to the form of the object) on the existing melamine board.

Example mold with sink or cooktop on it

When you're done, you can now drill pilot holes then join the mold together using dry-wall screws (length depending on the thickness of the countertop). Clean the bottom of the mold to make it as smooth as possible before pouring the concrete. Make sure your mold is supported well before pouring the concrete.

Use silicone to make the edges smooth

Once the sides have been screwed into place, it's time to apply the silicone caulk. This is to make the edges smooth for safety reasons and esthetics.
Use a gun to press the silicone (100% silicone) on the edges of the sides at the bottom and on the vertical corners of the form.

Use your fingers to spread the silicone evenly throughout the form. I believe some people are asking "is this process really necessary?" YES, adding silicone has a number of benefits such as sealing the joints together and also preventing wet concrete from penetrating through the joints. After applying, let the silicone dry for about 24 hours before wiping the inside with olive oil.

Applying silicone

Purpose of using olive oil

Now that the frame has been screwed together and silicone has been applied on the edges of the sides, it's time that you apply olive oil on the surface of the melamine board to make it smooth and most importantly, to prevent the concrete from sticking on the surface of the melamine.

This step is very important when preparing a DIY concrete countertop because it makes it easier to remove the sides and the bottom form when the concrete has dried up. Don't use a lot! A small coat is enough.

Reinforcing the concrete

Why should you reinforce?

Concrete countertops are very different from floors and other slabs because while the later (*floors & slabs*) are supported by a subgrade, the latter (*countertops*) are like cantilevers and need additional support to avoid cracking. The types of reinforcing materials you'll need to use are equally important as the type of mixing ingredients you'll need to use when making your countertop. Before we get to the reinforcing materials, let's first understand the physics behind reinforcing.

A concrete countertop is much like a cantilever therefore when a force is applied at the top, two forces; the compression (*on the top*) and the tension (*at the bottom*) will act on the slab.

Although concrete is known to have high compressive strength (*ranging from 3,000psi or 200bar*), it has very low tensile strength (*about 400psi or 30bar*) making it prone to damage such as cracking when excess force is applied at the top surface.

Did you know that a woman of around 110lbs (50kg) who wears a high heel produces 4300psi (300bar)? The tensile forces usually occur on a straight line along the slab (*not across*). Therefore, to ensure that your concrete is strong enough and capable of enduring the tension force at the bottom, strong reinforcing materials will need to be added and placed along the length of the slab. Just my little disclaimer: don't walk with high heels on your countertop. Why would you?

For added strength, people add reinforcing materials to ensure the tensile strength is equally distributed.

Now that you've understood everything about compressive and tensile strengths, we are now going to focus on the different type of materials used to reinforce a concrete countertop.

Using wire mesh
Ladder wires and mesh wires with enough steel have proven to work perfectly well when it comes to reinforcing concrete countertops. Most contractors prefer using mesh/ladder wire because they're flexible and capable of reinforcing simple concrete slabs with complex cutouts.

For much heavier concrete slabs, these wires can be combined together to provide a stronger reinforcing material that is capable of keeping the slab strong and tough. However, one drawback of using mesh wires is that they're only limited to rectangular concrete slabs and not molded designs.

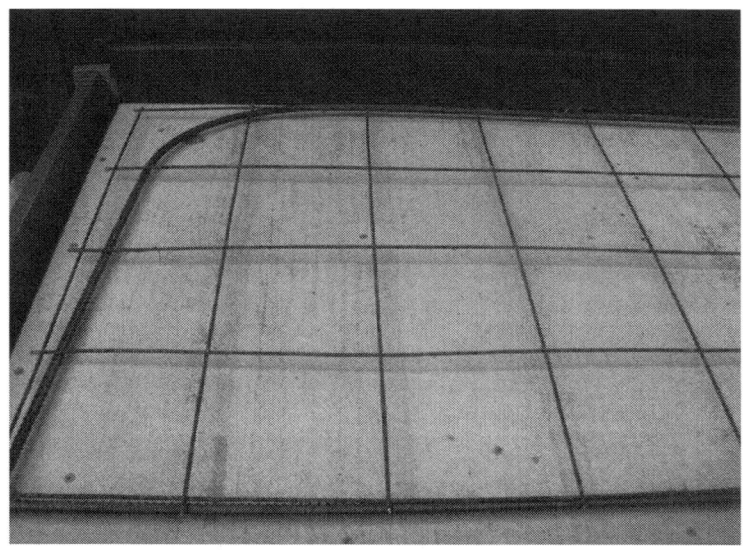

Rebar with wire mesh

Using Rebar

Most people prefer using steel rebar to reinforce concrete slabs. What happens is that 3/8 inch (1cm) rebar grids are tied together with 3/16 (0.5cm) pencil bars using thin wires they're positioned along the concrete.

Always make sure the larger rebar grids are placed along the concrete slab while the smaller pencil bars are positioned in a perpendicular angle to the rebar in every 6 inches (15cm). This type of positioning of the rebar and the pencil bars ensures that your concrete countertop remains strong for a very long time.

However, according to research, most professionals have warned people on using rebar as a concrete reinforcing option stating that it's too big for a normal sized concrete countertop. To prove this, take example of a normal sized countertop of either 1.5" or 2" (3.8 or 5cm). When the slab is reinforced with a 3/8 (1cm) rebar grid running along the concrete, you'll notice that the oversized rebar will occupy much space on the thin concrete slab.

This means that there will be less space left between the rebar reinforcement and the surface of the concrete. Although the resultant concrete countertop will have a great tensile strength of about 80%, the oversized steel rebar will cause a large stress on the small concrete covering the surface of your countertop leading to *cracking*.

Cracking of a countertop

Using carbon fiber grid
A corrosion resistant material that is a perfect substitution of both mesh wire and steel rebar is carbon fiber mesh. Carbon fiber is light, thin, flexible and above all, it has a great tensile strength.

A single membrane of carbon fiber is estimated to be 1/32 (0.08cm) inches and has a breaking strength of nearly 250 pounds (115kg). Carbon fiber is also very flexible making it perfect for binding tight and profile spots such as round curves, around the sink holes and areas at the top of the slab.

Coloring the concrete

There are many methods which homeowners use to color and decorate their concrete countertops. These methods change the color and appearance of a concrete countertop making it beautiful and stylish. Among the common methods used to color concrete include; using integral pigments, using dyes, using acid and stains and finally using decorative aggregates such as crushed glass and colored stones.

Using integral pigments

The use of integral pigments in coloring concrete is among the most common methods used by most DIY'ers. These pigments are mainly made from finely ground particles that are mostly mixed with the concrete during the mixing process to change its color. Integral pigments may be liquid or powdery in nature and they mostly come in a variety of colors such as blue, grey, green, brown, white, yellow or red.

When mixing integral pigments with your concrete, a lot of care must be taken especially when measuring the ratio of water. Through careful batching, mixing and curing, the resulting color consistency is usually (if you did it right) uniform. When the concrete is fully dry, you'll find that the color of your countertop is exactly what you expected without any errors on it.

When using the integral pigments method to color your countertop, always remember that these pigments come in different prices depending with the quality. In addition to this, you need to know that concrete is a harsh environment for integral pigments because its reaction with sunlight can attack the pigments breaking them down thus causing damage to the coloration. Therefore to avoid this, you need to buy the right pigment which is suited for concrete.

Finally, it's wise to understand that integral pigments come in a variety of different types which include;

- **Iron oxide pigments**: these are the most common type of pigments used to color concrete countertops. They are usually cheap, accessible and come with basic colors which include; grey, red, yellow and brown.

- **Synthetic pigments**: these pigments are more expensive than iron oxides and metal oxides because they're more powerful and vibrant. people are advised to use a small amount when mixing to get the required colors.

- **Liquid and powdered pigments**: liquid and powdered pigments are very common in most stores because they're the most used pigments when coloring a concrete countertop. Powdered pigments are sold in cans or bags and are mostly dustless because the particles are held together by water soluble binders.

    These pigments require one to use an electric mixer so that the particles can be crushed and mixed vigorously to break down and produce the required colors.

Using dyes

The second method of coloring concrete countertops is by use of dyes. Dyes are translucent, soluble and penetrating solutions that form a vibrant topical layer of color that is applied on the surface of your countertop. Dyes can either be water or solvent soluble and one advantage of using them over pigments is that they don't react with the concrete to damage the surface of your countertop.

Dyes are available in a range of different colors depending with your preference. They can duplicate the colors of acid stains or they can also take the colors of integral pigments. There are many advantages of using dyes over all other coloring techniques. Most contractors prefer using this method because it saves on time making it ideal for projects that require a fast turnaround time.

Although dyes provide great versatility when used to color and design specific areas on your concrete countertop, they are limited on where they're applied. Most dyes are UV unstable and can react with the UV rays to damage the coloration as well as the concrete countertop. To avoid this, contractors or homeowners are advised to apply the dye with either aerosol sprayer or acetone sprayer. To achieve this, apply at list three or four coats of your preferred dye color to achieve a darker and intense color. After you're done, seal the finish with either aerosol or acetone to protect it against wear or unwanted pealing.

Finally, let me say that coloring your countertop with dyes saves time and money. Although this method is cheap, the skill of applying is quite hard. There are lots of things you need to adhere to before applying the dye finish such as cleaning the concrete top and removing any dirt, mud, debris, paint, adhesives or anything that prevent the dye colors from bonding with the surface of your concrete countertop.

Using acid stains

Acid stain is another common method used by people to color concrete. Unlike using dyes and integral pigments, using acid stains to color your concrete countertop delivers professional results when done correctly. The resulting color is usually stylish, complementing and very solid meaning it can't wear off even when exposed to heat or direct UV rays. Acid stains are chemical solutions which when mixed with concrete, they react with the available calcium hydroxide (*in the concrete*) to produce color (*stain*). Inside the acid stains, there are metallic salts which are responsible for the reaction. The acid on the other side catalyzes the reaction resulting in the formation of particular stains (*colors*).

Using the acid stains method is very dangerous at some point because the results are usually spontaneous and unpredictable. However, there are a number of considerations one has to keep in mind before mixing the concrete for much better results. First, mix the ingredients in their correct ratio and ensure the texture of your concrete is smooth without any debris or adhesives. Make sure the acid stain solution is concentrated and finally, give the acid stain enough time to react with the concrete. To explain this section further, allow me to highlight a simple step-by-step procedure on how to color a concrete countertop using acid stains.

1. Firstly, clean your countertop thoroughly through wiping and removing any dirt, debris or residue. Proper cleaning of the countertop allows the acid stain to coat evenly leading to quality results.

2. Cover all the cabinets, sink and close walls with plastic then tape them securely.

3. Wear protective gear such as goggles, gloves and face mask; then open all the windows (if you are inside) for good aeration. This is to avoid the fumes from harming you when mixing the acid stain.

4. Pour the stain on the top of your countertop and spread it evenly using a paint brush.

5. Allow the stain to dry for at list three hours. You can apply additional coats if you're not comfortable with the first one. You'll notice that after every coating, the color of your concrete countertop becomes darker (if you are using black of course).

6. Wait for 24 hours then clean the countertop thoroughly with a clean sponge. After every round, you'll see some residue leftovers of acid stains. Remove every bit of them then leave the countertop to dry thoroughly.

7. Use an acid stain neutralizer or mix 2 cups of ammonia with 5 gallons of warm water to wash the top of the countertop. Wash it thoroughly to allow it set the original color of the concrete. When done, clean the countertop with clean water to remove any neutralizer present on your concrete. Allow the countertop to dry before applying a coat of epoxy penetrating sealer. This

helps the countertop to become waterproof and prevent the acid stain from peeling off (*remember the acid stain penetrates about a thousandth of an inch into the concrete so the color can easily wear off due to abrasion*). Allow the countertop to dry for another 24 hours before using it officially.

Acid stains are very economical to use. A gallon costing about $50 can paint a very large section of concrete. Mixing the acid stains together can deliver many different colors which can be used to paint a bathroom or a kitchen concrete countertop to match the color of your home. This method is simple, economical, fast and of course unpredictable.

Using decorative aggregates

Unlike in the past, people are becoming more creative through finding ways of improving the appearance of your kitchen or bathroom concrete countertop.

They're actually coming up with cool ideas and entirely new methods of coloring your countertops to make them more decorative. One of these modern ways of decorating a countertop is through the use of decorative aggregates such as pieces of glass, mica, bits of mirror, pozzolans and precious stones. To help you understand more about the various decorative aggregates, let's take a look at them:

Use of glass aggregates

Use of glass aggregates to decorate concrete countertops is among the newest methods currently preferred by most contractors.

There are special companies which deal with recycling and reshaping of different types of glass into different sized cubical shaped chips which are used for decoration. These glass recycling companies use damaged glass from windows, bottles, mirrors and other glass products

Glass aggregates are available in a range of 20 different colors. There are two colors which are the most expensive and are quite difficult to get. These colors are red and orange. Another way is to simply make your own by crushing some glass bottles. Although I do not advise this for safety reasons.

There are two major methods that can be used to apply glass aggregates. One can either use the seeding method where glass particles are placed inside a form before concrete is added or the mix all method where the decorative aggregates are mixed together with concrete during the mixing process.

Rock salt finish
The rock salt finish is a traditional and easy method of decorating a concrete slab to make it rough and skid resistant. This method gives your countertop an old vintage appearance without any weathering effects.

When creating the salt finish, simply press the coarse salt rocks onto the surface of the concrete before it's fully dried. You can use a roller to press the rocks uniformly to avoid damaging the surface of your concrete (*make sure the rocks protrude halfway above the surface of the slab*). When you're done, leave the slab to dry for about 24 hours before power washing/brushing the salt off. You'll notice that after the salt particles are washed off, speckled patterns of different sizes will be revealed making your concrete countertop a bit rough and stylish.

One drawback to using this method is that it can only be done while your concrete is wet. You can only do this when you use the cast in place method as seen in the first chapter.

Rock salt finish

Mixing the concrete

Concrete mixes

From kitchen countertops to bathrooms and fireplaces, the art of decorating countertops has pushed the limits of concrete. While there are so many creative ways of decorating concrete countertops, the success behind everything begins with proper mixing of the concrete. There are two basic methods which most people use to make high quality countertops. They include;

- Using bagged mixes
- Making the concrete mix yourself

Bagged mixes
Bagged mixes are perfect for making concrete because they arrive ready and packed with the necessary ingredients to accomplish the mixing task.

However, this method is quite limited as it doesn't offer you a chance to alter the mix through adding any ingredient apart from what's in the bag. You therefore need to trust the manufacturer's ingredients and instructions when adding water and mixing. Due to this reason, most people prefer to go for the do it yourself mixing method because it gives you the freedom to add your own decorative and strengthening ingredients to make your concrete countertop strong, solid and stylish.

Let me say that concrete countertop mixes are being manufactured by so many companies today. These countertop mixes arrive in the market when they're mixed in the correct ratios giving you a simple task of just adding water before mixing. According to experts who have tested the strength of several concrete countertops, there are several mixes which have proved to perform a perfect task of designing strong and smooth countertops. They are;

- **Quikrete 80lb Commercial Grade concrete countertop**

The Quikrete Commercial Grade concrete countertop mix is the best concrete mix used to design concrete countertops. This concrete mix is designed with smart features that make mixing, pouring and vibrating the concrete fast and easy. It features a super-plasticizer additive that requires you to use just a small amount of water. This concrete mix is further designed to make it very strong, smooth, easy to polish and paint with any color of your choice. The Quikrete Commercial Grade concrete countertop mix takes only 18 hours to completely dry and is capable of reducing shrinkage for a long-lasting use.

- **enCOUNTER professional countertop mix**

The enCOUNTER professional countertop mix is a concrete mix designed with a blend of high enhancing ingredients, fine graded sand and a blend of high quality Portland cement. When mixed according to the right measurements, the enCOUNTER has the capacity to design strong, durable and heat resistant concrete countertops. Unlike other concrete mixes, enCOUNTER has the capacity to design countertops with less shrinking, cracking and curling drawbacks. Finally, this concrete countertop mix has a high comprehension strength that makes it resistant to wear and tear.

- **Cheng concrete countertop mix**

The Cheng concrete countertop mix is another reliable concrete mix that's highly recommended for beginners of DIY concrete countertops. Designed with great strength, this concrete countertop mix is made with strengthening agents that enable it to design smooth, strong and crack resistant countertops. This countertop mix supports fibers and a variety of colors such as stone, wine, amber, olive, ocean, base, indigo, charcoal, evergreen and jade. We will talk about coloring in the next chapter.

Do it yourself mixing

One reason why most people use this method is because it gives you unlimited control over your work. Although people are advised to use bagged mixes because the manufacturer has already mixed the ingredients in correct ratio, those who understand the mix design and how to measure the ingredients are advised to use the from-scratch mix method.

The first step of making a strong concrete countertop is by understanding the water-to-cement ratio (w/c). Note that most of the characteristics of concrete such as the strength, porosity, shrinkage, color, durability and curling tendency are directly dictated and controlled by water. As you all know, concrete becomes harder and stronger when less water is used during the mixing process. Although some people add more water to improve workability, it's good to understand that you're likely to suffer from tough consequences.

- The entire structure is likely to become weak and porous.
- The decorative crystals will keep apart and unable to knit together making the decoration unprofessional.
- The decorative colors will appear pale and more washed-out.

Therefore, to avoid suffering from any of the three consequences above, people are advised to use superplasticizers instead of water when mixing concrete from scratch.

Superplasticizers are special chemical admixtures (*also referred to as high range water reducers*) which are added to concrete mixes to make them smooth, creamy and pourable. The reason why these chemical admixtures are used is because they increase workability when preparing concrete mixes while maintaining the concrete's strength and toughness.

Mixing the concrete

Pouring the concrete

Where to pour the concrete?

After the mixing process in a bucket, wheelbarrow or on the floor, it's time to pour your concrete. There are two methods. They can either use the mold method where the concrete is poured on a well-prepared mold or they can pour it directly on the countertop (cast in place). To help you understand this better, let's review each of these methods.

Pouring concrete on a mold

This is one of the most common methods used to design concrete countertops. It involves pouring the concrete into an already made mold. With the use of a hand trowel or a float, spread the concrete throughout the mold until the entire area is fully covered. Ensure that you've set the mesh wire or mixed the carbon fiber within the concrete. Use a hand trowel to smooth the top surface of the concrete.

Pouring concrete directly on top

Another common method that is being preferred by most homeowners is by pouring the concrete directly on to the countertop.

To successfully achieve this, you can decide to use a new melamine board or you can simply pour the concrete on the existing kitchen countertop. If you decide to use a completely new board, make sure that you take the correct measurements of the existing countertop.

Also, make sure that you pay special attention to any cutouts such as PVC pipes, sinks and cooktops. After you've taken all the necessary measurements, construct wooden forms and screw them together all around the kitchen countertop.

Clean off any debris or dirt particles on the surface of the form to ensure that the concrete stays smooth and straight. Apply silicone or tape the edges of the forms to ensure that wet concrete doesn't penetrate through when it's poured.

Measure the length and width of the available fiberglass/carbon fiber mesh and fit it precisely inside the forms. After everything is set, you can now mix the concrete and pour it gently. Use a hand trowel to spread the concrete uniformly until the entire countertop is filled up. Use a magnesium or steel float to smoothen the top of the concrete countertop until it's uniformly smooth and appealing.

Vibrating the mold to remove air

After the fresh concrete is poured onto the mold/countertop mold, the next step is to use a hand trowel to smoothen the top. After you've done that, you'll now need to use a vibrator to remove any air bubbles that could be trapped inside the concrete.

Note that, air bubbles should be removed at all cost because failure to do so may leave the surface of the concrete countertop with unattractive holes or pits. However, not all concrete needs vibration. Some concrete mixes come with gentle ingredients which tackle this problem automatically. On the other hand, some homeowners may prefer working on a rough slab especially if it's meant for outdoor use.

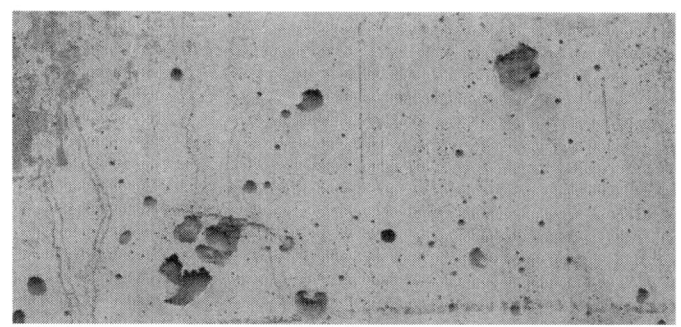

Result of not removing air bubbles

What causes the air bubbles?
Most people have been wondering what exactly causes air bubbles. Bubbles occur when air is trapped between the melamine board and the concrete during the pouring process. This air is known as entrapped air and it consists of large air bubbles. In addition to this, if the concrete is not mixed thoroughly, small air bubbles are likely to get trapped in the concrete mix causing bubbles. This air is known as entrained air.

Therefore, to avoid the concrete from producing air bubbles, you have to mix it thoroughly, ensure its flowable/workable then use a vibrator *(maybe a hammer or an orbital sander)* to tap or vibrate the mold until all the air bubbles are released. When you start doing this you will see the air bubbles burst open at the top of the mold.

If the air bubbles stop popping, then you know you are ready. If there still are some airholes when you turn the countertop, you can add some concrete on top to seal the holes.

(Removing air bubbles leaves your concrete countertop smooth and free from holes which hold debris, dirt and food particles making it difficult to wipe the countertop).

Screed the top

Immediately after you pour the concrete and vibrate the mold to remove the bubbles, it's time to screed the top of the concrete slab. The main reason for screeding the concrete is just to level it and make sure that all the high and low spots are removed. If you are using the mold technique, you will level the bottom of your slab.

If you use the cast in place technique you will level the top. Screeding is somehow a very challenging process. This is because you'll have to battle with time where the process must be accomplished before the concrete gets dry and stiff.

How to let the concrete slab dry slowly

As soon as the concrete slab is smoothened with a trowel or has been leveled, its' time to allow the concrete dry slowly by itself. This process is known as curing. For a concrete countertop to gain strength, hardness and density, moisture must be retained inside the concrete for a continuous and progressive hydration. When cast concrete is kept moist, it continues to gain strength and lose porosity making it hard and very solid at the long last.

During concrete curing, proper temperatures ranging between 50°F to 75°F (10 – 25°C) should be maintained. If the temperatures are too low or too high, the curing process can either be curtailed or accelerated leading to weak results in the long run. To ensure that your concrete slab remains strong, you have two options which include using plastic sheets or Membrane-Forming Curing Compounds.

- **Membrane-Forming Curing Compounds**
  These are chemical compounds that are directly sprayed on the surface of a cast concrete to form a thin blanket of waterproof membrane that prevents moisture from penetrating through to the environment.

- **Plastic sheeting**

    This method is the most common in most concrete countertop processes. It involves covering the top of the newly cast concrete with a plastic sheet to prevent moisture from going away too fast. Always remember to cover the top surface gently to avoid damaging the countertop surface. Lifting the plastic sheeting using some tools or scrap wood is a good idea. Try to make it look like a tent over the concrete slab.

Finishing the countertop

Ways to check if it's dry

Testing concrete countertops for excess moisture has become a common practice in this modern world of construction. This step is very important because it allows you to determine the best time to smooth or color your countertop with paint or epoxy. To help you understand more, this section will review some of the basic and affordable methods used to check for excess moisture in a concrete slab. Standard drying time is around 4 days.

Use of clear plastic
One of the best ways of determining the moisture levels of a concrete slab is through taping a small piece of clear plastic on the surface of the slab. Leave it for 16 hours then check if you can see any moisture presence at the bottom of the plastic. It there is moisture it needs more curing.

Calcium chloride
The use of calcium chloride is regarded as one of the oldest and most preferred method of testing for concrete dryness. This test is done with special kits which are left on the surface of the concrete to determine how much moisture is coming from the slab. This is the more expensive option.

Relative humidity testing kit
This is the only sure way of testing whether a concrete is fully dried or not. It involves inserting RH probes inside small drilled holes on the surface of your concrete to measure humidity levels. This is an expensive method too.

I recommend sticking to the use of clear plastic because of costs and simplicity. If you are not sure just leave it to dry for a few more days.

Turning the countertop over

Now that your concrete slab has fully dried, it's now time to strip the melamine sides from the mold. First remove the screws then use a trowel to gently strip the wooden sides without damaging the concrete.

Use an orbital sander or a grinder to smooth the sides of the concrete slab before gathering a few friends to help you flip it over. Note that, depending with the thickness of the slab, concrete countertops can weigh up to 200 pounds (90kg).

You therefore need to take extra care for your back and let some friends come over to help flip the countertop. Then, finish the day of with your friends and some cold beers.

Protecting the countertop

Concrete is vulnerable to certain food acids such as fruit juices, vinegar, wines and mustard. Apart from that, it's highly vulnerable to the daily wear and tear caused by dragging of cook pots and pans from one place to the other. To ensure that your countertop is well protected, it needs to be sealed immediately after drying up. In this section, we will highlight ways in which you can protect your concrete countertops from damage.

Wet polishing

There are so many benefits of polishing a concrete countertop after it has cured. However, the main reasons why polishing is necessary is to make the top surface pleasingly smooth while removing any blemishes, dirt or debris particles. In addition to that, polishing a concrete slab exposes some bug holes which would not have been visible during the early stages.

When preparing for the polishing process, you need a hand-held wet concrete polisher, diamond grinding pads with different grit levels and safety goggles and ear plugs.

Start off by washing the concrete surface with water to remove all grit and dirt particles. In case of any holes, fill them first then allow them to dry before proceeding. Start the polishing process with the coarsest grits depending with the level of smoothness you wish to achieve. Use the hand-held diamond blocks or pads to polish profile areas such as round corners and on the edges where electric polishers can't reach. Allow the concrete to self-dry naturally before applying your desired food safe sealer.

Wet polishing with a grinder

Penetrating sealer and concrete sealer

Just like I explained earlier, bare concrete is vulnerable to various chemical attacks and stains from different types of foods and drinks. These stains contain concentrated acids that dissolve to the concrete to react with the cement paste to damage the appearance of the concrete countertop. Therefore, sealing the concrete surface provides additional protection to the concrete preventing these stains from damaging the surface top.

Note that, there are two types of sealers available for concrete countertops. They include, penetrating sealers and concrete sealers.

Penetrating sealers
Penetrating sealers are among the most common sealing methods used to protect concrete countertops. These sealers are mostly used to seal concrete on coffee table tops, bathroom and fireplace concrete tops where less activities takes place (*as compared to the kitchen*).

The reason for this is because these sealers don't offer a long-lasting protection to aggressive staining agents. Using penetrating sealers on your kitchen countertop gives you a task of cleaning your kitchen top every now and then to avoid any damage caused by stains.

Concrete sealers

- **Acrylics**
  These are some of the commonly used concrete sealers in the market today. Available in either solvent based or water based sealers, acrylics usually dry up to form a hard and protective layer on the surface of your concrete. The hard layer is both water and UV resistant making it perfect for "busy" kitchens. However, acrylics have one major setback. They can easily be scratched leaving the surface exposed and prone to damage. They therefore need frequent reapplication to make them stronger.

- **Epoxies**
  These are other concrete sealers that are very common in the concrete market. When mixed, epoxies chemically react to form a smooth, hard and durable seal on the surface of your countertop. The reaction is usually rapid meaning you have to apply it pretty fast to avoid any mess. Epoxy is available in three basic types which includes; 100% solid based, solvent based and water based. There are several disadvantages that come with epoxy though. One, this sealer is highly vulnerable to UV exposure meaning it can be chemically broken down when exposed to direct sun. Two, because it forms a thick hard seal, it can be scratched easily leaving the surface exposed and prone to damage.

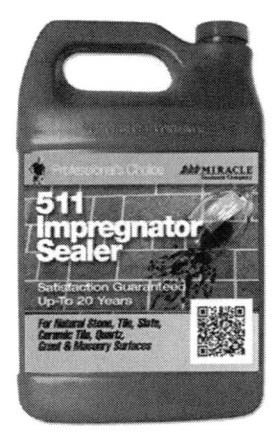

Miracle Sealants for concrete countertops

How many coats?

When spraying concrete sealer to your countertop, you need to check the thickness of the sealer to make sure it's precise. The number of coats you'll need to paint depend with the recommended thickness of the sealer you'll have to achieve. Most manufacturers provide customers with a wet mil gauge which helps you to measure the recommended thickness of the sealer. Note that, in case the sealer is too light, it doesn't offer the expected protection. On the other hand, if it's too thick, you have done an unnecessary cost.

Is cutting food on countertops toxic?

Okay, let me say that this depends with the types of materials used to make the concrete countertop. The best thing about it is that most manufacturers who deal with the production of concrete ingredients are coming up with more eco-friendly ingredients which are perfect for making non-toxic green concrete countertops. In addition to this, manufacturers of concrete sealers are also coming up with food-safe sealers which are non-reactive when fruits and other foods are directly exposed to the countertop. Therefore, people should first research on the best eco-friendly concrete mixes and sealers based on the availability in local stores.

Final words

Conclusion

To recap I will go over the steps you need to take to make your own concrete countertop. This countertop.

1. Measure your desired dimensions for the countertop.

2. Pick up some melamine board from your DIY store and make the frame. Do not forget to add an overhang.

3. Check if there are any faucets or kitchen sinks. If there are, make your frame accordingly.

4. Finish the inside edges with silicone. After it is dry add some olive oil to the melamine board to prevent it from sticking.

5. Begin with adding your steel rebar or mesh for extra strength. Elevate this because you don't want to see the rebar sticking out of your countertop.

6. Mix your concrete and add colours if necessary.

7. Pour the concrete in the mold and shake or hammer the air bubbles out.

8. Screed the top to remove excess concrete.

9. Let the concrete dry for some days with plastic on top.

10. Remove the mold carefully and flip it over with some help.

11. Inspect the top of your countertop and fill in the holes (if there are) with some concrete you have left (use the same ratios for colour otherwise you will have spots).
12. Let it dry.

13. Start to wet polish your countertop for a smooth surface.

14. Seal the countertop.

15. Place the countertop on its final place and work the edges with silicone.

Thank you

Since concrete is the primary material behind most building foundations, making kitchen countertops from this material makes it undoubtedly the best in terms of strength and quality.

I would like to thank you again for purchasing this book. If you learned anything from this book (which I'm sure you did), Please leave a five-star rating on amazon and recommend this book to your friends and family. I would highly appreciate it!

Thank you and take care!

Bradley Specter

Made in the USA
Coppell, TX
09 September 2020